INDEX

THE BOOK ABOUT EVERYTHING

Professor J.T. Springston

God Equation; $(1A + 1B) + (1AB) = \{1C\}$

JT' Excalibur; "The group is the byproduct of the individual components and is only then itself one individual component of a higher dimensional grouping".

Extrapolating Dichotomy Logic;

Because $(1A...1B)$ / (Birth + Death)

Because $(1A + 1B)...(1AB)$ / (Birth + Death) + (Life)

Because $(1A + 1B) + (1AB)...\{1C\}$ / (Birth+Death) + (Life) $= \{Lifespan\}$

Parameter Limits. Higher Dimensional Individual Component. Repeat.

EXAMPLES;

1. (Point A + Point B) + (Line AB) = {String 1C}

2. (Beginning + End) + (Middle) = {Timeline}

3. (Birth + Death) + (Life) = {Lifespan}

4. (Yes + No) + (Maybe) = {Choice}

5. (Right + Wrong) + (Ambiguous) = {Moral}

6. (Win + Lose) + (Draw) = {Outcome}

7. (Up Quark + Down Quark) + (Color Quark / Gluon) = {SubAtom}

8. (Neutron + Proton) + (Electron) = {Atom}

9. (Hetero-nuclear + Homo-nuclear) + (Poly-atomic) = {Molecule}

10. (Bacteria + Archaea) + (Eukaryotic) = {Cell}

11. (Up + Down) + (Level) = {Height}

12. (Left + Right) + (Center) = {Width}

13. (Height + Width) + (Breadth) = {Area}

14. (Egg + Sperm) + (Womb) = {Baby}

15. (Boy + Girl) + (Hermaphrodite) = {Gender}

16.(Divergent+Convergent)+(Transform)={Tectonic Plate Boundaries}

17. (Classical + Dwarf) + (Satellite) = {Planet}

18. (Transmit + Receive) + (Range) = {Communication}

19. (Conscious + Unconscious) + (Subconscious) = {Awareness}

20. (Off + On) + (Standby) = {Power}

21. (Red + Yellow) + (Blue) = {Hue}

22. (White + Black) + (Gray) = {Contrast}

23. (Hue + Contrast) + (Composite) = {Color}

24. (Small + Large) + (Medium) = {Size}

25. (Inhale + Exhale) + (Hold) = {Breathing}

26. (Heavy + Light) + (Average) = {Weight}

27. (Space + Matter) + (Energy) = {Existence}

28. (Father + Son) + (Holy Spirit) = {God}

CHAPTER 1: Introductions Are In Order

Dr. Kaku once refered to the ToE Equation as a "basic fundamental structure". I refer to it as a "universal blueprint". Communication is the hardest obstacle to overcome in this world. Transmission needs reception. There's a whole range between the two wherein everything can be lost, misinterpreted, wasted. How the human species has managed to make it this far is a testament to the power of that communication. If I have ever failed at anything on this journey it was in finding the right ways to reach people to explain the data on their terms.

There is the complex, the intricate, the expansive nature of logic and how it can become overwhelming to people but there is also the simple, the elegant, the easy to to process. It is in those areas that I work diligently to reverse engineer from "super complicated science thingy" down to "well yeah, that seems pretty obvious". The problem lies in convincing your average person that they are far more intellectually capable than they realize. That the basics of science have always and will always remain the same; OBSERVATIONAL DATA.

It's what we do as a species. We collect data. Then we use it to get more data. First you realize that there is fire. Then you realize that you can make it. You realize that you can do a better job of making it. Then you realize what you can use it for elsewhere. You make other things out of the fire that you made. It's all extrapolative logic.

Don't overthink it or you'll miss the forest through the trees. Everything in life is just engineering and reverse engineering in a state of constant evolution. Where we error more often than not in Physics is getting lost in what I refer to as "Singularity Logic"; the focus of breaking everything down to or building everything up to a singularity and thereafter discarding the existence of the previous components.

What does that mean in simple terms? The Judaic and Christian divide is a good example. The Jews treat God as a singularity. The Christians/Catholics, etc. follow the Holy Trinity of God, the individual components. My work involves putting these philosophical principles together to always acknowledge both the collective results and the individual components that they are comprised of.

As Dr. Kaku has dubbed this "The God Equation", which he and I agree on, most others in the scientific community treat it as the "Holy Grail" of science and statements on religion will be somewhat unavoidable in this book. It will be limited to assess all of this work from a strictly scientific lens. I accept that there are those within the scientific community arguing that to discuss science is to discuss religion and I agree. I'd ask for some leniency from all sides as per my descriptions. I will try to be delicate and respectful with language throughout this book, no matter your particular belief structrure. To be clear, I am Jewish. I was not always.

That said, reference the equation from page 1. What you will notice if you compare to any other work regarding a ToE Equation is that my equation not only has a RULE but also a RESULT. To compare to Dr.Kaku's Hyperspace Equation, "Height+Width+Breadth+Time", it is unfortunately and immediately flawed because of this. He had no generated result. Be it "E=MC2" or "A2 + B2 = C2", there absolutely must be a generated result. I do respect his work, he was on to something. His failures lied in a few simple flaws that extrapolated into our field work coupled with a redundancy in Geometry/Physics logic matrices as he applied them.

I feel that in fairness to Dr. Kaku and his lifetime of work dedicated to the ToE Equation, he deserves much respect. I am also left to critique his work as there are very few people to this point that have offered anything tangible to dissect. That is to say that his books and collective concepts all stood out amongst the mass amount of frivolous pseudo-science floating around, parading itself as a "potentiality", without so much as a simple equation or description, that doesn't involve 3 degrees to understand the terminology. Thereby, completely refuting itself as applicable work relating to an equation that is rooted in simplicity and the very nature of logic.

Still, I am left to dissect his work as an unfortunate byproduct of the state of science, which I see as stagnant and in need of inspiration. To do so, I am obligated to present my own work in advance or I would truly be leaving you lost in the woods. I will be making the comparisons as you read throughout, with a recap and overview. As per now, let's keep it simple by addressing the fluidity of evolution and how it applies to the human fallacy logic that I've dubbed as "Singularity Logic".

Singularity Logic is a mental trap that has lead to many a human being being lost in a rabbit hole for good. Don't go down it...

But if you are down it, you should really be reading this. One apple is just one apple.

We can argue over Platonism and Bertrand Russels' views on Metaphysics as we go for those who will, most undoubtedly, consider them almost immediately when presented with relatable subject matter. I digress, that apple contains many seeds. Those seeds might all grow to become trees themselves, so on and so forth to infinity.

Some philosophical perspectives focus only on the notion of "oneness" and the idea that since all of the trees came from one apple that they all then must be from the "one" apple tree. This leading to the idea that a forest is just "one" tree if they're all the same type of tree. It spits in the face of evolution in every regard. Not to be outdone is the notion that nothing is connected and everything is just randomness. Without wasting too much of my time on that, we'll just call it a bit nonsensical, which will be proven and then, move on.

The God Equation or ToE(Theory of Everything) accepts the other ToE(Theory of Evolution) and therefore "plays by the rules" of evolution. Answering the age old question "Which came first, the chicken or the egg"? It was the Egg and yes, it matters. At some point a chicken egg was laid by a species that was not a chicken. The egg contains a chicken. By deciding that you're discussing the chicken you have set the parameter limits at the species before the chicken, which laid the egg that hatched the chicken. The question answers itself.

Time and evolution work hand in hand. The previous species continues on its' own evolutionary path and the new species begins a tangential variant of the previous. This logic is even acknowledged and represented in the same spiritual divide between the Christians and the Jews when referencing The New Testament quote from Jesus that "It would be easier to move every star in the sky than to change one stroke of the pen of The Old Testament".

The past can not be altered. Whatever species came before, remains. Order of operations is inherently; Space first, Matter second, as per engineering in physics. A new evolutionary tangent does not erase the existing timeline of another species.

Thus the avoidance of "Singularity Fallacy" and the notion of "oneness" in any tangible sense. You can not apply "Infinity" to everything and claim that there is no such thing as an individual because of it. Or to then use representative things as "realisms" that lead to the Platonism that muddies the waters of the question "what is existence"

In short, if you'd like to talk about what doesn't exist, feel free to discuss with someone that doesn't exist and leave the rest of us alone.

CHAPTER 2: Greek Dichotomy Logic Explained

(& Its' Application To (1A+1b) + (2b+2A) Mirrored 2D-4D Geometry)

We've all been taught similar logic; "If there is Life... then there is Death". Simple and easy to understand. It's also flawed as applied. That statement in and of itself is not a dichotomy. It's a "Middle" and an "End". Where is the "Birth", I.E. the "Beginning"? This is a BASIC, FOUNDATIONAL, STRUCTURE that is applied to all of our field work. It's the spawn of flawed conclusionary data in multiple fields because it is the foundational principle of all of our logic. Even "Yes and No" are the building block of "binary coding" as pertaining to computers; 1's and 0's.

If Beginning; Then End

If They; Then Middle

If Those; Then Timeline

If (A; Then B)

If (A + B); Then (AB);

If (A + B) + (AB); Then {1C}

Thus the Equation; $(1A + 1B) + (1AB) = \{1C\}$

10

Applied per "Life/Death" Logic it repairs the matrice as applied on a field level. To such an extent that it actually proves Mirrored 2D-4D Geometry flawed as applied to 3D Physics. It is itself a 3D Geometrical equation that applies to 3D Physics. As such;

If Birth; Then Death .... If A; Then B

If (Birth + Death); Then (Life) .... If (A + B); Then (AB);

If They; Then Lifespan ..... If (A + B) + (AB); Then {1C}

Thus the Equation ..... $(1A + 1B) + (1AB) = \{1C\}$

Note that both of these equations are identical. They acknowledge both the individual components and the collective results of them. This is the proper format and order of operations that actually exists in our reality as applied to all fields. I can not stress enough the overwhelming abundance of evidence that exists to validate my claims and the inability to prove myself wrong in spite of my best efforts.

In a word, the work is simply "True". I will give examples below and throughout, with an index of all examples given at the end of the book. Some of which will also be used as examples of "Stretched ToE", which will be addressed in a later chapter.

(Beginning + End) + (Middle) = {Timeline}

(Birth + Death) + (Life) = {Lifespan}

(Egg + Sperm) + (Womb) = {Baby}

(Proton + Neutron) + (Electron) = {Atom}

(Height + Width) + (Breadth) = {Area}

(Space + Matter) + (Energy) = {Existence}

(Father + Son) + (Holy Spirit) = {God}

No matter the field, subject matter, etc., the same equation is observed. It is how logic itself extrapolates. There is a Primary Dichotomy. Because it exists there is a Secondary Dichotomy. Because it exists there is a Third Dichotomy. As I will illustrate later in the chapter on "Stretched ToE" I intend to prove that those are the parameter limits and that the equation extrapolates back into itself, hence the { } when describing the collective result. Any "Atom" is just one atom that can be either a cation, anion or isotope. Any "Baby" is just one baby that can be either a boy, girl or hermaphrodite. As it extrapolates it stretches into a format of;

(Rule + Result) + (Representation) = {Collective}

By proxy turning right back into the original equation of;

(1A + 1B) + (1AB) = {1C}

This observation proves Mirrored 2D-4D Geometry flawed as applied to 3D Physics for this reason; wether engineering or reverse engineering in any field, be it an "atom" or an "area", everything breaks down or builds up into 3 dimensions. Fusions are independent dimensions. They are a 3rd by proxy dimension in any field that counters the primary dichotomy, creating the secondary dichotomy, then the third. It is because of this that Newtons' work pertaining to the "Laws of Motion" need be amended.

Newtons' 2nd Law of Motion is a Law of how Energy itself functions. He made an unintended error in his description during a time period that predates what we now understand about fusion by roughly 400+ years. He completely discounts fusion as possible as per the wording. It needs an amendment.

Also, I feel people need a reminder of the beginning of the statement; "When in the Absence of External Forces", which is generally disregarded when considered in other work. This is the rule of "perfect world", as all rules should be. I will make the argument that the only thing in this world that is perfect is the rules, which I see as reality, which is to say that we don't always get the rules correct but that the rules do not change.

In fact, I would argue that the best that I, or anyone can ever do as an educator, is to help others to better understand the reality that exists. I have changed nothing with my observations other than my understanding of the world around me. Consider this equation a long overdue software update for your brain that improves understanding of the foundational reality logic that exist of its' own accord.

13

As per Newtons' 2nd Law of Motion and the application of my ToE Equation, as stated previously "Stretched ToE" will be explained further in the book but this is a prime example of how well the logic extrapolates and how good the equation is at correcting errors.

$$\text{(Unstoppable Force + Immovable Object) + (Absence Of External Forces)} = \{\text{Yield}\}$$

Based on the order of operations, you could only have an "Absence Of External Forces" if "Internal Forces" are present. This is the same equation of $(1A + 1B) + (1AB) = \{1C\}$. How does it stretch and how does it correct Newton? His "Either/Or" yield prospect completely discounts fusion as one of the Result options. When you stretch the String (ToE Equation), you always have 1 of 3 "Or" options present. In Result this means that;

$$\{\text{Yield}\} = \text{(Fusion) OR (Force Yield OR Object Yield)}$$

$$\{1C\} = (1ab) \text{ OR } (1a \text{ OR } 1b)$$

Newton intendted to prove that "Yield Will Occur" and he did just that. In stating that "One Must Yield To The Other" he discounted the fusion option as per the result of how energy functions. You can see the error on a field level when applied to Physics.

14

In essence, we know that HERMAPHRODITES exist in Biology. They are not 2 babies, they are 1 baby, because they are a fusion result option. The ISOTOPE is not 2 atoms. It is 1 atom, positive and negative primary dichotomy result options must also account for their counterpart, the 3rd dimension.

Applied to Biology you see the same trend; (Egg + Sperm) + (Womb) = {Baby}. The womb does not activate until a zygote is present; primary dichotomy of sperm and egg. It is the 3rd dimension Rule option that only exists because the primary dichotomy exists. Comparable to the "Absense of External Forces" in its' (1AB) placement. The Baby then having 1/3 result options upon creation i.e.;

$${Baby}=(Hermaphrodite)OR(Boy\ OR\ Girl)$$

$${1C}=(1ab)OR(1a\ OR\ 1b)$$

When expanding your equation remains the same. Intent matters. What are you looking for? Do you want to know how to make a BABY, what a BABY is after it is made or all of the collective data surrounding the BABY?

This does not alter the equation, it naturally extrapolates on a dimensional level from 1D-2D-3D. This equation is Super Symmetry & it proves that Super Symmetry makes the rules for engineering from a lower dimension up to a higher and vice versa. When putting Rules and Results together your equations will then look like this;

15

***   {

$$(1A + 1B) + (1AB)$$

$$= \{1C\} =$$

$$(1ab) \mid (1a \mid 1b)$$

}

***   {

$$(\text{Unstoppable Force} + \text{Immovable Object}) + (\text{Absence External Forces})$$

$$= \{\text{Yield}\} =$$

$$(\text{fusion}) \mid (\text{force yield} \mid \text{object yield})$$

}

***   {

$$(\text{Egg} + \text{Sperm}) + (\text{Womb})$$

$$= \{\text{Baby}\} =$$

$$(\text{hermaphrodite}) \mid (\text{boy} \mid \text{girl})$$

}

In closing, I have concluded that Newtons' 2nd Law of Motion need be amended to "Yield Will Occur" from "One Must Yield To The Other" in order to account for fusion. This is in no way a slight to Newton or his any of his work.

This is merely an acknowledgement that his final summary was ill worded, which had unintended consequences as it applied to all work regarding energy in Physics. In order to move forward in the realm of Physics and as per how energy functions, this amendment need be accepted and considered in all of our work moving forward.

There are other implications across various fields that I have partially touched on above regarding Biology specifically. Like any foundational problem it has severe consequences as you move higher up the engineering chain. Things get more and more unstable. Nuclear fusion is a great example.

Our current methods have us using improper Geometry applied to Physics leading us to require more energy to start reactors than they can themselves generate. This is the equivalent to needing a generator to start your generator. This issue can be, need be and will be resolved if I have a say so within my lifetime I can assure you.

Reference The Zeroeth Law of Thermodynamics- "2 Thermal Systems in Equilibrium with a 3rd System are in Equilibrium with each other". This requires not only an inherently independent 3rd dimension as per equilibriums in the vague "thermodynamics" sense but refutes using a 2D-4D push method as specifically applied to nuclear fusion. We need to be using a 3D "Pull" Method that will naturally draw the energy towards fusion and no, I will not be publishing my designs in this book. I will be publishing more data in the future as I finalize projects, etc.

CHAPTER 3: String Theory & Stretched ToE

(Successes & Failures Of String Theory As Understood & Applied)

"If you want to find the secrets of the universe, think in terms of energy, frequency and vibration" - Nikola Tesla

Many hands have put a paint brush to the artwork that is "String" or "String Field" Theory and/or Quantum Mechanics. Dr.Kaku, Stephen Hawking, Gabriel Veneziano, Leanord Susskind, Brian Greene, Einstein, I give credit to even Nikola Tesla for his consideration of particles and waves.

The comparison to the String is easy to visualize. Until now, no one has been able to properly explain how it functions in basic terms. Some mistakes lied in a misapplication of 2D-4D Geometry in a 3D environment. Others rooted in Einsteins' Light Observations as applied to Time.

Many errors have been made that revealed themselves more and more as the natural course of engineering was taking place. The better your logic, the higher you can build on an engineering level before things start to topple over.

This is where we're ended up as pertaining to String Theory. I now have a proper description of how it functions and why. It stretches with the data because of the natural extrapolation of Greek Dichotomy logic. Some of you will understand that statement immediately, others may need to do some study to understand the full depth of that statement. String Theory itself applies on an individual field level and engineers up. The String Stretches or shrinks with the data. Let's go back to the beginning;

"If it exist its' equal opposite exist". This is your Point A and your Point B. As referenced in Chapter 1, that's your Beginning and your End. Because they exist? The line exists, AB, the Middle. You can't have a Middle without both a Beginning and an End. This is all based on the same logic currently used. I only claim to have a greater understanding of it. As such;

If you have 2 end points (1A + 1B), their equal opposite is then the line between them (1AB). Extrapolating further, if you have those individual components (1A + 1B) + (1AB), you have their equal opposite, the constructed "String" {1C}. This all collectively forming the equation;

$$(1A+1B)+(1AB)=\{1C\}$$

You can extrapolate data and expand the string but the equation then turns back into itself on a higher dimension, which is to state that you are just going from D1 to D2 or from D2 to D3 as you exapnd your data. The equation remains the same even as you expand. That sounds far more complicated to people than it actually is in my experience.

I'll do my best to simplify my description with a few examples on individual field levels. Find a field that you're familiar with and use it as a guide;

*****STRETCHED ToE*****

$$(1A + 1B) + (1AB) = \{1C\}$$

$$\{ (1A + 1B) + (1AB) = \{1C\} = (1ab) \,|\, (1a|1b) \}$$

BIOLOGICAL ToE Stretched

$$(Egg + Sperm) + (Womb) = \textbf{\textit{\{Baby\}}}$$

$$\{ (Egg + Sperm) + (Womb) = \textbf{\textit{\{Baby\}}} = (hermaphrodite) \,|\, (boy \,|\, girl) \}$$

CHEMICAL ToE Stretched

(Neutron + Proton) + (Electron) = *{Atom}*

{(Neutron + Proton) + (Electron) =*{Atom}*= (isotope) | (anion | cation)}

GEOMETRICAL ToE Stretched

(Height + Width) + (Breadth) = *{Area}*

{ (Height + Width) + (Breadth) = *{Area}* = (pi) | (parameter | plot) }

ENERGY ToE Stretched

(Im Object + Un Force) + (Absence External Forces) = *{ Yield }*

{ (ImObj + UnForce) + (Absence External Force) = *{Yield}*

= (fusion) | (object Y | force Y) }

RADIO ToE Stretched

(Transmission + Reception) + (Range) = *{Communication}*

{ (Transmission + Reception) + (Range) = *{Communication}*

= (wavelength) | (am | fm) }

When comparing and contrasting the stretched individual field equations you should notice the same trends as you follow the engineering order of operations;

- Rules are OBJECTIVE (AND) options represented by a (+)

- Results are SUBJECTIVE (OR) options represented by a (|)

- Representations are ABSTRACT (AND/OR) collectives represented by a { }

When stretching the equation your original {1C} Collective ToE Result beomes the (1AB) Abstract Representation between the (1A Rule / 1B Result) options.

The idea goes back to the same Greek Dichotomy Logic that this all stems from; "If it exist, its equal opposite must exist". I meant it when I've previously stated that it continues to extrapolate. It's not "complicated" but "expansive" and as you expand you maintain your dimensional parameters. You've gathered more data but now you have the same equation via;

(1A Rule + 1B Result) + (1AB Representation) = **{1C Collective}**

If a Rule exist then a Result must exist.

If a Rule and a Result exist, then a Representation of them must exist.

If the Individual Components exist then a Collective of them must exist.

The same logic that is used in the beginning stages of ToE is the same logic expressed once you stretch a particular subject matter to gather more data. It's all "If;Then" logic. Having already addressed Newton as pertaining to "Stretched" ToE I want to address the Geometry of "Stretched" ToE and how it applies to Pi. First, let's look at the primary dichotomies and how they function. They create a singularity and they are countered by a singularity. This can be easily understood when looked at on individual field levels. Examples;

(Neutron + Proton) = (Nucleus)

(Egg + Sperm) = (Zygote)

(Transmission + Reception) = (Signal)

This then leading to the secondary dichotomy;

(Nucleus) + (Electron) = {*Atom*}

(Zygote) + (Womb) = {*Baby*}

(Signal) + (Range) = {*Communication*}

Note the order of operations apparent in each field. The Electron is the Energy dimension of the atom that moves around the nucleus as a counterforce. The Womb is only activated once a zygote is present, until then it is only a Uterus. It is an autonomous dimension activated by the existence of the zygote. I would postulate that pertaining to the yielding of shield walls, with the amendment I've made to Newtons' 2nd Law of Motion to account for fusion, the Electrons are created because of the fusion of the Nuclei. It would follow the same order of operations that seems to apply in every field of Physics.

Pertaining to Geometry the process is identical; (Height + Width) + (Breadth) = {Area}. The primary dichotomy is first; (Height + Width) create a plot. That is their fusion. This process has a byproxy result; the Breadth. In a 2D Geometrical environment, drawing the (Height + Width) forms a (Plot) the same way that the (Egg + Sperm) form the (Zygote). There is a 3rd dimension brought to life in the same way as the Uterus is activated into a Womb upon the presence of a Zygote; (Breadth).

Breadth is known without drawing it simply because the primary Height and Width lines exist. Their existence let's you know the parameter limits without drawing them. It's comparable to the 3rd dimensions in every other field of physics. Energy can be measured without being seen. The Breadth is known without being drawn. The Womb is activated without being forced. All of these act as by proxy reactions to the fusion of the primary dichotomies, thus the extrapolating nature of "If it exist, its' equal opposite must exist".

Imagine that you start with a string in any field because of the Greek Dichotomy Logic. Because there are 2 end points, there is the singular connecting string. Again, to reference Thermodynamics "2 Thermal Systems In Equilbrium With A 3rd System Are In Equilibrium With Each Other". These are basic, fundamental, rules at play on a multi field level. I'm not "making things fit" my work. All I've done is expand upon the logic that we already knew to be true and corrected the errors wherein we've applied that logic poorly.

25

CHAPTER 4: Identifying Infinity & Quantum

(How The Chemical Chain Is Built & Properly Establishing
Parameters)

*"If You Think You Understand Quantum Mechanics, You Don't
Understand Quantum Mechanics" - Richard P. Feynman*

To respond in my defense; If I don't understand Quantum
Mechanics then no one does. Apply Stretched ToE;

$$\{ (1A + 1B) + (1AB) = \{1C\} = (1ab) \mid (1a \mid 1b) \}$$

$$\{(\text{Neutron} + \text{Proton}) + (\text{Electron}) = \{\textbf{Atom}\} = (\text{isotope}) \mid (\text{anion} \mid \text{cation})\}$$

To break down the Proton/Neutron into their quark patterns is
to enter into an INTERNAL QUANTUM of the "{Atom}". You are no
longer within the parameters of the subject matter "{Atom}" and have
now altered the parameters focal point to the "{Subatom}";

$$\{(\text{UP Quark} + \text{DOWN Quark}) + (\text{Gluon}) = \{\textbf{Subatom}\}$$

$$= (\text{electron}) \mid (\text{proton} \mid \text{neutron}) \}$$

26

To build up from atoms to a "{Molecule}" is to enter into the EXTERNAL QUANTUM of the "{Atom}". You are no longer within the parameters of the subject matter "{Atom}" and have now altered the subject matters focal point to the "{Molecule}";

$$\{ (Anion + Cation) + (Istotope) = \{\textbf{Molecule}\}$$

$$= (polyatomic) \mid (heteronuclear \mid homonuclear) \}$$

To build up from molecules to a "{Cell}" is to enter into the EXTERNAL QUANTUM of the "{Molecule}". You are no longer within the parameters of the subject matter "{Molecule}" and have now altered the subject matters focal point to "{Cell}";

$$\{ (Heteronuclear + Homonuclear) + (Polyatomic) = \{\textbf{Cell}\}$$

$$= (eukaryotic) \mid (bacteria \mid archaea) \}$$

Naturally occuring engineering pattern creates an overlap, forming a chain;

{

(UP Quark + DOWN Quark) + (Gluon)

= {Subatom} =

(Neutron + Proton) + (Electron)

= {Atom} =

(Anion + Cation) + (Isotope)

= {Molecule} =

(Heteronuclear + Homonuclear) + (Polyatomic)

= {Cell} =

(Bacteria + Archaea) + (Eukaryotic)

}

A good mathematical representation of this pattern can be found when reviewing Fibbonaci Sequence. Engineering and reverse engineering are both step by step operations in either direction, at their maximum potential. This chemical chain is the exact representation of my Razor;

"The Group Is The Byproduct Of The Individual & Is Only Then Itself One Individual Component Of A Higher Dimensional Grouping". This is the natural state of evolution. People make up Neighborhoods. They then make a City. Cities make up States that compose Countries. Those Countries are within Continents. Those Continents are all a part of a Planet that is a part of a Solar System.

The Solar Systems then make Galaxies that then make up Universes. Universes make up Multiverses and I would postulate that then there would be a "Cluster" of Multiverses. It can all be engineered and reverse engineered down to the Quark and up to the Multiverses.

My best advice for any intellectual involved in science is to always remember the fluidity of evolution and time, without getting lost in the "Infinity" of it all. Stating that everything is all just one forest doesn't really help much when you need an apple. It's not all just "the same". There is variance.

There are equal opposites, then self equal opposites, etc. The masculine man has the feminine man. The feminine woman has the masculine woman. The proton has the anti proton and the electron has the positron.

This is equation is how engineering and reverse engineering is done on a basic, multi-field level. It is a universal blueprint for how all of the field work functions. This is because it is all based on the Greek Dichotomy that got us here. It solves some of the greatest issues we deal with in Physics to date. It also validates a very important rule that we are all aware of;

"Every rule must have an exception" is itself a rule. It requires an exception to be valid. My ToE equation is a rule without exception. Thus, it is the exception to the rule that "Every rule must have an exception", thereby validating it. Until now, there was not one. The only exception that could exist is the reality rule. The absolute truth of how the world around us functions.

The only "Perfect" in life is the rule. Perfection has to be reverse engineered as compared to the Rule. The rule of making babies is that a man and a woman have sex. Yet, they can have sex numerous times without ever making a baby. This does not mean that the rule is "wrong". What it does mean is that one of the components is not at its' maximum potential. The sperm can be less than perfect. The egg can be less than perfect. The womb can be less than perfect. As a matter of fact, that is the most likely outcome in all pregnancy. True perfection can only exist in rule, not in result.

Expand your parameters. Presume that a baby is made. What kind of baby can it be? The options are Boy, Girl or Hermaphrodite. Still, these are representative of "Perfect" options. They are not a real world result. Even though the options are limited, these options all exist on a spectrum. Apply the same logic to any field. You can say yes, no or maybe to something. Each of them can be responded with varying levels of enthusiasm, reasoning for doing so, etc. There will always exist variance. Variances and spectrums do not alter the RULES. They merely expand upon the data that exists.

Newtons' 2nd Law of Motion actually validates this prospect in spite of its flaw in discounting fusion. The entire premise is based on "Perfect" equal and opposite forces that do not exist. An unstoppable force is to this point unknown. The same is true for the Immovable Object. How could a Theory using theoretical infinites be proven as a

30

Law? Because Rules are not Results. Rules are the only things that are perfect. These forces are meant to be representative of perfect world outcomes. We then observe the data available to us in real life and compare it to the rules. That's how science works. It's a constant state of evolution with a set of underlying, fundamentals that don't change.

Apply the same logic to Communication. (Transmission + Reception) is "Perfect" transmission and reception. The Wavelength or Amp or Frequency that the Communication generates are all "Perfect" in rule. What we deal with in real life can be quite different. We know all about static and interference and any number of issues that cause a transmission to not be perfect when it arrives and to be less perfect when broken down or rebroadcasted. Rules are meant to be perfect, not reality.

When dealing with terms like "Quantum" and "Infinity" you have to keep these things in mind to avoid falling into "Singularity" fallacies. Einsteins Light Observations are a good example of "losing the parameters". Those observations were then applied to Time as an absolute in physics. The study of white light is the study of the "Infinity" variant as pertaining to Light. The combination of all light options on the spectrum in its' truest form. Of course if you applied it to time it would treat time as "Infinite".

This led to another flaw in Dr.Kaku's Hyperspace sans the application of 2D-4D Geometry in a 3D Physics environment; considering Time both Geometrical and Material and treating it as inclusive of matter. All because, and this is no joke, Einstein studied white light and it was later determined that that was how "Time"

31

functions. As I've previously established in this book, time is fluid if there is to be any truth to evolution at all. It does not disprove the components of Time; (Beginning + Middle) + (End) = {Timeline}. Focusing on the collective of all timelines and grouping them into one timeline as an infinite completely disregards basic evolutionary principles.

Dr. Kaku treated Time as a 4th Geometrical Dimension that was inclusive of Matter. Somehow making a fusion of Time and Matter into one field dimension that is also a geometrical dimension. I apologize for seeming crude but it makes absolutely no sense and its' why it was universally refuted within a short period of time. My format makes more sense;

Stretched ToE Applied;

$$\{(Area + Atom) + (Energy) = \{\textbf{Existence}\} = (Time) \mid (Space \mid Matter)\}$$

Basic ToE;

$$(Area + Atom) + (Energy) = \{\textbf{Existence}\}$$

Area breaking down into Geometrical ToE, Atom into Chemical ToE, etc.

My ToE equation is a 3D Equation meant to be used in 3D Physics. It proves that 2D-4D Geometry is the wrong tool for 3D Physics and properly formats the universe into 3 independent result dimensions of Space, Matter & Time. As applied to the individual fields;

Biology: Space = Egg / Matter = Sperm / Time = Womb {Existence/Baby}

Chemistry: Space = Neutron / Matter = Proton / Time = Electron {Existence/Atom}

Radio: Space = Reception / Matter = Transmission / Time = Range {Existence/Communication}

Holy Trinity: Space = Father / Matter = Son / Time = Spirit {Existence/God}

CHAPTER 5: Banach-Tarski Is No Paradox

(Geometrical Trickery & Ill Applied Logic)

"Mathematics is the queen of science & Arithmetic the queen of Mathematics"- Carl F. Gauss

This mathematical proposition has been debated for quite some time and I refute it as a falsehood. There are not more numbers in between 2 numbers than the numerical timeline that they are broken down from. Banach-Tarski is not a paradox. Reference points are #1 and #2. You are not between them. You are passed the one. You are on the right side of the decimal.

— Time moves forward, never backward

— Electrons move clockwise, never counter

— Numerical timelines move to the right

What this means is that you are only dealing with pieces of the 2 as the predetermined next number in sequence. The 1 is finished. It is whole. You are now dealing with the next whole number in sequence and breaking it down. To add digits to it is the same as pertaining to PI, this is simply a ZOOM function. You are expanding on a single piece, not multiple pieces. For example;

If I pick a plot point passed the 1 but before the 2, I am only picking a point of the 2, thereby a quantity of the 2. It is a plot of a numerical line that is representative of a partial quantity of the 2nd digit. To add digits to it? Apply to weight. Instead of 1.11 lbs you want to be more exact, so you go to the next digit possible, 1.111 lbs.

To continue to add digits does not alter the quantity sans choosing to round up or down, which we know is in and of itself false data as we've determined the next number in sequence already and decided that we do not need it as per the desired data comparisons. You are only using a more precise metric. You are simply extracting more data.

As per Banach-Tarski, it expresses the same expansion of data collection as is apparent in other mathematical calculations. There is no paradox. To add digits passed a plot is to expand the data you have to a broader view. The original data isn't altered but expanded upon.

Expansion of data is not paradoxical. There is no "greater than" infinity that is "in between" the two whole numbers. All that is presented here is the ability to break a whole number down into smaller components to gather more exacting data. Following the same order of operations the mysteries of Pi reveal themselves. Pi is a Fusion Result of (Plot+Parameter);

{

(Height + Width) + (Breadth)

= {AREA} =

(Pi) | (Parameter | Plot)

}

Height and Width first make the Plot. Because THEY; IT. They are the reason for the existence of the Breadth. Breadth is an automatic by proxy occurrence. It is known without drawing it. The individual components then creating the collective {AREA}.

Area/Diameter requires that you find the Hypotenuse to solve... That's the halfway point. The reason for this is because the plot and the parameter are in a symbiosis with one another. Their relationship is entangled automatically. Making Pi a fusion Result option. Field Reference Radio;

Result options are AM or FM. But they are both broken down from a Wavelength. Wavelength is a fusion result option. As is the Hermaphrodite. As per Geometry? Pi is no different.

Adding digits to it is equivalent to zoom on a camera. You are on the right side of the decimal. You are not altering the whole. Your 10ths, 100ths, 1000ths metric is comparable to 10X, 100X, 1000X. It is a ZOOM function; An expansion of the plot.

CHAPTER 6: Proving Godel' Proof Flawed

(The Study Of; "No Set Of Axioms Can Prove Its Own Consistency")

Zero is not a number. 0 is the symbol for the complete absence of a quantity. There is also no 10 in a base 10 numerical system. This is verified with a repeating mathematical pattern, discovered by writing out time tables from 1×1 through 1×30 to 30×1 through 30×30, then reducing your sums to singularities by adding them together again and again until only 1 digit remains;

1 × 12 = 12 = 1 + 2 = 3 .... **1 × 13** = 13 = 1 + 3 = 4 .... **1 × 14** = 14 = 1 + 4 = 5

10 × 12 = 120 = 1 + 2 + 0 = 3

10 × 13 = 130 = 1 + 3 + 0 = 4

10 × 14 = 140 = 1 + 4 + 0 = 5

The pattern for the number 10 is the same pattern for the number 1. 1-9 follow a previously unrealized, distinct set pattern, when reduced to a singularity. Once you get to multiples of 10? The patterns repeat. Endlessly. You see this again when you reach 19;

19 × 1 = 19 = 1 + 9 = 10 = 1 + 0 = 1

19 × 2 = 38 = 3 + 8 = 11 = 1 + 1 = 2

19 × 3 = 57 = 5 + 7 = 12 = 1 + 2 = 3

37

Again at 28;

28 × 1 = 28 = 2+8 = 10 = 1+0 = 1

28 × 2 = 56 = 5+6 = 11 = 1+1 = 2

28 × 3 = 84 = 8+4 = 12 = 1+2 = 3

By expanding to greater sums with larger numbers that I was able to verify that this is a pattern with purpose.

"Zero" is the equal opposite to "Infinite". Zero is the absence of a quantity, not a quantity in and of itself. Which is co-validated by not being present anywhere in a repeating pattern where any quantity actually exists.

Thereafter studying the quantification of the base 10 numerical system itself, I noticed that there is no 10 pattern. Nor a 35 pattern. Nor any pattern that did not stem from one of the already existing patterns created by 1-9.

Every sum passed the 9 follows a pattern of 1-9 that can be discovered via adding it's own digits together in the beginning to find its' pattern WITHOUT a need to actually write out all the times tables and reduce the sum to a singularity. With any number, without the times tables, you can determine its' pattern by adding itself together;

18 follows the 9's pattern (8 + 1)

64 follows the 1's pattern (6 + 4 = 10 = 1 + 0 = 1)

0 × Anything = Always Itself

Notice 9's pattern here;

9 × 1 = 9

9 × 2 = 18 =1 + 8= 9

9 × 3 = 27 =2 + 7= 9

9 × Anything = Always Itself

It is a repeating, byproduct, pattern of the Base 10 numerical system, that follows naturally occurring dichotomy logic in numerical engineering. The base 10 numerical system is a mathematically balanced logic matrix of equal opposites.

1-9 are the only real numbers, provable with an infinite, repeating, pattern. Zero (0) is not a number. It is a symbolization for the absence of any quantity. Whereas Nine (9) is equal opposite to the zero (0) as both always equal themselves in result.

When you write out all of the tie tables adding your sums to singularities the patterns are as follows;

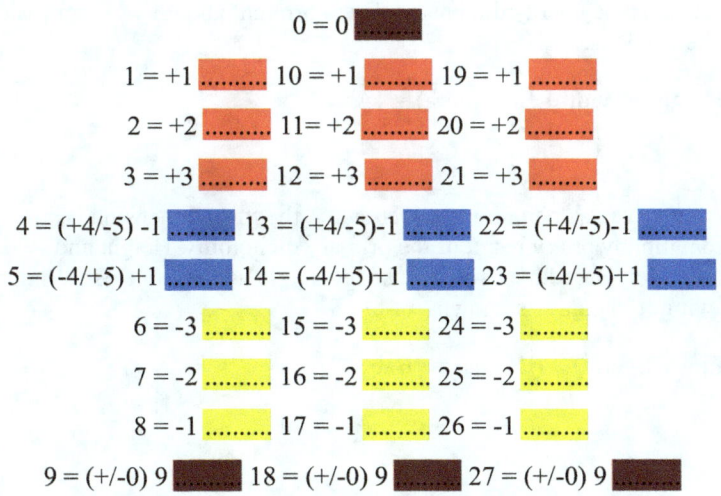

0 = 0
1 = +1   10 = +1   19 = +1
2 = +2   11= +2   20 = +2
3 = +3   12 = +3   21 = +3
4 = (+4/-5) -1   13 = (+4/-5)-1   22 = (+4/-5)-1
5 = (-4/+5) +1   14 = (-4/+5)+1   23 = (-4/+5)+1
6 = -3   15 = -3   24 = -3
7 = -2   16 = -2   25 = -2
8 = -1   17 = -1   26 = -1
9 = (+/-0) 9   18 = (+/-0) 9   27 = (+/-0) 9

39

CHAPTER 7: Life Begins At Conception

(Order Of Operations & Autonomous Biological Function)

If Zygote; Then Womb. A womb is just a Uterus until a Zygote is present. The human female does not decide to form a womb. The body does of it's own accord. The body chooses to feed the baby. The woman can not starve the baby, only herself, therein by proxy the baby.

As such, I have come to the following conclusion, following the same scientific format of "(Space + Matter) + (Energy) = {Existence}" so apparent on a multi-field level in that;

- (Proton+Neutron)+(Electron)={Atom}

(+1 + +/-0) + (-1) Balanced Forces from naturally extrapolated Dichotomy Logic. Because Nucleus; Valence. The valence shell only exists because there was a proper fusion of (Neutron+Proton). The Electron is the energy dimension that is brought about by proper fusion of the Nucleus.

- (Height + Width) + (Breadth) = {Area}

Even as pertaining to Geometry, the third dimension is an automatic, by proxy result of the primary dichotomy. Height and Width form a Plot. Breadth exists because the height and width exist. Without drawing it, the data is known.

- (Egg + Sperm) + (Uterus) = {Baby}

Because Zygote; Womb. Egg and Sperm for a minimum fusion potential. Thereafter, the Womb is activated by the Zygote the moment that the Zygote has fused inside of the Uterus. It is not only by proxy but also autonomous.

Therefore, no matter the field, the same format and order of operations apply. The same rules of "by proxy" and the implications thereof. The autonomous nature of 3rd dimensions.

The woman's body AUTOMATICALLY takes over once a zygote is present, without any input from anyone. It is an autonomous 3rd dimension that only functions because the primary dichotomy exists and thereafter, automatically functions without any further initiation.

Once a zygote is present and it has activated a womb? Life has begun. The autonomous energy dimension has sprung to life. To argue against that is to argue against science.

We all start off as Hermaphrodites. There is not a Baby that does not need Both an Egg and a Sperm. You need both parts so that we all start off as Hermaphrodites. We GROW into one direction or the other via Newton's Laws on Force Yield. When you properly account for Fusion you reveal important data for the engineering order of operations on a field level.

So that a Baby will never be a "Perfect Boy", "Perfect Girl" or a "Perfect Hermaphrodite" even. They are unattainable things but they are all 3 necessary to account for in Result. Hermaphrodites are not an "Exception". 3rd options are not "Exceptions". They are the compromise dimension. The Hermaphrodite is the singular, fusion, compromise dimension of the Boy/Girl dichotomy.

41

$$(Sperm + Egg) + (Womb) = \{Baby\}$$

And then...

$$\{Baby\} = (Hermaphrodite) \mid (Boy \mid Girl)$$

We START as Hermaphrodites. The hermaphrodite is the "Fusion Result Option" that occurs when both dichotomy options yield to each other so that, when everything yields? Nothing does. That's why it's a compromise. Dichotomy logic is at play.

We GROW into Boys and Girls throughout the 9 month process with meta force yield data that is genuine but intricate. The amount of underlying factors are staggering to consider but they exist.

Dichotomy Logic needs to be extrapolated to fully comprehend Fusion as the singular duality that automatically exists as an option by proxy the fact that the Primary Dichotomies themselves exist. If you follow the format? The Womb appears 3rd in Rule. The Hermaphrodite appears 1st in Result.

The engineering logic is flawless. There is an equal and opposite balance from Rule to Result when accounting for 3rd dimensions. This follows function in Thermodynamics and Newton when you accept FUSION in Newton' work, which he accidentally discounts as impossible with bad verbiage.

CHAPTER 8: Unattainable Perfection

(Validating The Rule Of Exception & Zombies & Unicorns: Thesis On
Evolutionary Potential)

*"The greatest desired result is achieved when everyone does what's
best for themselves & then the group" - Dr. Jonathan Nash*

Every rule must have an exception" is itself a rule. It requires
an exception to be valid. The exception to that rule is a rule with no
exceptions. It validates the entire premise. My ToE equation has no
exceptions, thereby making it the exception to the rule that "Every rule
must have an exception". The rules themselves are also "perfect".

Rule 1: Nothing in this world is Perfect.

Rule 2: Every Rule must have an Exception.

Rule 3: A Rule Can Not Be Its Own Exception.

*****Rule 1*****; "Nothing in this world is Perfect".

-Exception-; The Rules themselves are Perfect.

Reference: Biology;

(Man+Woman)+(Sex)={Baby} is a Perfect Rule. I.E. Failure
at Reproduction is not failure of the Rule. The rules address the
maximum potential of perfect world systems. So that the Rule itself is
(100% Man + 100% Woman) + (100% Sex) = {100% Baby}. In reality,
this is the one Result that you are guaranteed to never attain. This is due
to the constant evolution and intersects in biological reproduction.

43

Reference: Newton's 2nd Law;

"Unstoppable Force" and "Immovable Object" are theoretical "Perfects". The Rule still applies to imperfect kinetic examples. Also forgotten by many is the beginning of the Law; "In The Absence Of External Forces". This is all meant to establish perfect world parameters. In reality all closed systems are constantly being manipulated by external systems. Consider human beings; It is harder to NOT have an impact on the world around us that it is to have an impact. This is seemingly never more apparent than as pertaining to "Butterfly Effect" logic.

Reference: Game Theory;

Even as pertaining to economics and mathematical matrices like Game Theory, the Rule requires a perfect world caveat; "All Things Being Equal". This is correct, although I find it to be flawed in application as it relies on everyone doing what is best for themselves. In reality, we know that people do not always do what is best for themselves.

Some Players sacrifice themselves for the benefit of the other Players. Some sacrifice the goal of winning for the pleasure of playing the Game rather than trying to win. Some players are inexperienced and do not know what is best for themselves. Game Theory is designed as a "Perfect World Rule" which is to say that if you apply it in reality it will generate flaws based on people violating the parameters of the Game itself. Game Theory relies on the notion that human beings are selfish and will do what's best for themselves as a primary function. The Rule is Perfect. The Reality is not. We reverse engineer perfection by comparing Reality to the Rules.

*****Rule 2*****; "Every Rule Must Have An Exception", must itself have an Exception. Remember that it can not be its own Exception. Up until now, we have known no such Exception. What this means is that not only is a ToE Equation possible but, it is also REQUIRED to validate the Rule that all "Rules must have an Exception".

44

Without a Rule that has no exception, which would then apply to literally everything, we had no validation for the Rule that Rules must have Exception. An exception to that Rule is required and therefore a ToE Equation is required for Rules to have to have exceptions to them. Otherwise all Rules could be exceptions to themselves thereafter, invalidating the Rule entirely with the logic; "If everyone is anything then none of it is".

-Exception-; The God Equation / ToE; The Rule Without Exception.

Conclusion; The God Equation / ToE is then the Rule without Exception that validates the Rule of Exceptions, that was until now without an Exception and therefore, it was INVALID. To continue to use it without acknowledging an exception is to ignore the prove/disprove scientific method. This was actually a primary concern when trying to validate the work. It appears flawed at first as something that can not be disproved and therefore, can not be proven. It is because of this that even to personally validate it took a lot of time. I had to be absolutely certain that it was in a word, in every way that it could be, FLAWLESS. Only then could it be logically accepted as a Rule with no Exception. It is the only Rule that does not require an exception to be valid and that had to be proven with thorough compare/contrast assessments across all fields.

Practical application via "Zombies and Unicorns" was my attempt to place practical logic into an impractical environment with representation that might be easier for some people to digest. It is in no way a statement that zombies or unicorns are ever going to exist. It is a statement on how to take control of evolution while acknowledging the dangers associated. All that said, I can not make a unicorn. This is a simple statement of fact. It is an impossibility. By any metric or definition it can not be done in reality. Now imagine that we live in a constant state of evolution and that for some odd reason, we were immortal or in the least, had so much time to waste that it was of little to no concern to us..

Understanding that, I also ca not make a dog have sex with a cow and get a Unicorn. That's impossible. But if we we're going to live forever or to such an extent that it was not a concern, then over time I could breed the dog with other animals. Then even more animals with the offspring. I can do the same with the cow and I can slowly over time, 1,000's to 1,000,000's of years even, make a kind of a dog and get it to have sex with a kind of a cow. Much later, somehow, someway, ,overtime, maybe if you throw in a Rhinoceros, you get a kind of a unicorn.

That is Reality compared to the Rules. No I can not make a unicorn but if you were to give me the time and the resources? I could potentially get very close. So close that there would hardly be any distinction in some cases. Apply that same logic elsewhere. Perfection is unattainable.

No, you can not make "zombies", that is an impossibility. However, if you give me enough time and I can breed this with that or maybe that with this? I can actually mix a kind of a "this" thing with a kind of a "that" thing and I can make kind of a "zombie". I could even get really close. We see some of this naturally occurring in nature which has been transferred into recent video games and movies that exaggerate it. Nonetheless, mad cow disease and other such viruses have existed for quite some time and they are very well known, including prion diseases. All of these have given a few scientists here and there concern for what might happen later on down the evolutionary road.

I'm not really worried about zombies. Under the right circumstances it seems like everybody loves to fantasize about them. I am however, very concerned about Virology and Chemistry or Biological Reproduction as pertaining to Evolution. These things are of much concern. I do not condone many of the studies that take place and would much prefer that the focus was on better data collection and processing in controllable environments.

Still, I do understand the desire to explore the data wherever possible to try to assist our bodies natural abilities rather than to try to subvert or replace them with co-dependency of any variety. Also is the concern for parlaying or transferring data on a field level from one to the other with bad extrapolations. I.E. Using biological data applied to virology. One positive of such is in the way of realizing that inbreeding is bad and not doing it to viruses.

Unfortunately, that Rule appears to have been forgotten as of late as pertaining to Virology Research. There are both positive and negative side effects of transferring data from one field to another, with most of the negative impacts reside within the practice of bad formatting or transference before any errors are made pertaining to extrapolations thereafter. Dichotomy Logic is the primary extrapolation method for data collection and due to the symbiotic relationship inherent to dichotomies, it is easy to make errors without a blueprint helping to navigate new data.

The ToE Equation is the "basic, fundamental, structure, of all things" as stated by Dr. Kaku. I refer to it as a "universal blueprint" that applies to any field. Logic requires that it must be the Exception to the Rule that every Rule must have an Exception. It helps to format any new data as you collect it. It can then be applied to the field specific data within the format as it applies to whatever "new" data that you're dealing with. This is a necessary tool to have for when you are expanding throughout the universe and coming into contact with unknown specimens. You simply pick a field to study the new material with; Chemistry, Biology, Mathematics, Etc.

The ToE equation blueprint is the ultimate compare/contrast rule for multi-field work. Everything follows the same blueprint. If you can understand the data in one field then you are able to make sense another field. Reference Biological Reproduction. Apply stretched ToE;

{

(Sperm + Egg) + (Womb)

= {Baby} =

(hermaphrodite) | (boy | girl)

}

The same exact format applies to communication;

{

(Transmission + Reception) + (Range)

= {Communication} =

(wavelength) | (AM | FM)

}

Compared?

Transmission	= Sperm
Reception	= Egg
Range	= Womb
Wavelength	= Hermaphrodite
Amperage	= Girl
Frequency	= Boy

Primary Dichotomy and Fusion options are accounted for in both Rule and Result.

The same blueprint applies to every field;

ToE;

$(1A + 1B) + (1AB) = \{1C\}$

ToE Stretched;

$\{ (1A + 1B) + (1AB) = \{1C\} = (1ab) \,|\, (1a \,|\, 1b) \}$

CHAPTER 9: P vs NP Solution

(Accounting For Fusion Via "Primary Dichotomy Logic" As Applied To Physics Using An Extrapolated 3D Equation To Solve.)

$$(1A + 1B) + (1AB) = \{1C\} = (1ab) \mid (1a \mid 1b)$$

This is coupled with my Razor(Excalibur), which states that;
"The group is the byproduct of the individual components and is itself an individual component of a higher dimensional grouping".

The solution then is actually rather simple. Full logic matrix;

Because P = Solvable (1A);

Then N = Unsolvable (1B);

Because (P + N) Easily Solvable/Not Easily Solvable, (1A + 1B);

Then (NP) Verifiable/Unverifiable (1AB)

1. If (P; Then (N

2. If (P + N); Then (NP)

3. If (P + N) + (NP); Then {PNP}

$$(P + N) + (NP) = \{PNP\} = (pn) \mid (n \mid p)$$

(

P = Easily solved problems

+

N = Not Easily Solved Problems

)

49

=

(NP Verifiable/Unverifiable options)

=

(pn unverified)

| OR

(p verified solved | OR n verified unsolved)

       Because of the existence of easily or not easily solved problems, the primary dichotomy options, there then exists the equal opposite to that dichotomy; verification of the dichotomy options. Cross Reference the Zeroth Law of Thermodynamics as a field reference. The 3rd systems in any equilibrium are engineering checks for the primary dichotomies. I.E. Biology; If you don't have a proper fusion of a Zygote? You won't get a Womb. Also Chemistry; If you don't get a proper fusion of a Nucleus then you don't get a Valence Shell. The 3rd systems check for minimum potential symbiosis of the primary dichotomy options.

Step 1 - Is it easily solvable or not? A or B.

Step 2 - If it is solvable or not, is it easy to verify? (AB)

Step 3 - There will be a Result. = {1C}

Step 4 - Which will exist as 1 of 3 options;

- Unverifiable

- Verifiable solved

- Verifiable unsolved

{1C} = (1ab) | (1a | 1b)

Summation;

(P) does not = (NP)

(P + N) = (NP)

        Verifiable and Unverifiable only exists because Easily Solvable and not Easily Solvable exist. If something can not be solved, easily or no, it can not be verified. Verification is secondary to solution. Once easily solved and verified or uneasily solved and verified, the verification can then be either that the information is;

(Verifiable True or False) or (Unverifiable)

        The primary dichotomy is a dual-singularity that is countered by a singular-dual option. They are equal opposite and co-exist. I.E. The Hermaphrodite is the fusion option of the Primary Dichotomy of Boy/Girl but it is its own, singular, dimension.

CHAPTER 10: Einstein Theory Of Relativity

(Einstein Stated That, If Wrong About General Relativity It Would Only Take One Scientist To Prove Him Wrong. I Am One Scientist.)

"What we perceive as the Force of Gravity arises from the curvature of Space-Time".

His work is improperly formatted. It's very simple. If Time is the Energy dimension and Einstein would not disagree, only that it is also geometrical and chemical but he missed the point, General Relativity itself then was flawed. Follow the rules of thermodynamics and Newton' Kinetic Forces;

1. "Two Thermal Systems In Equilibrium With A Third System Are In Equilibrium With Each Other" - Zeroth Law of Thermodynamics on Equilibrium. The basics are how all Equilibrium function.

- Applied to Chemistry it means that you don't get a Valence shell unless there is a minimum fusion potential reached between the Proton/Neutron. This would also mean that all work around "One Electron Theory" is flawed elsewhere, based on the order of engineering operations in this field.

- Applied to Biology it means that you don't get a Womb unless you have a minimum fusion potential reached between the Sperm/Egg. This would mean that Life begins at conception based on the order of engineering operations in this field.

(Note that the Electron is the one of the 3 in motion. The nucleus maintains a fixed position. The Electron only moves Clockwise and never Counter-Clockwise in the same way that Time only moves forwards and never backwards. The same applies to Biology wherein you can not "Unmake" a Baby in a Womb, you can only "Make" a baby)

- Applied to Geometry it means that you don't get a Breadth unless you have a proper minimum fusion potential reached between the Height/Width. Meaning that they have to intersect at some point.

- Applied to Communication it means that you don't get a Range unless you have a proper minimum fusion potential reached between the Transmission/Reception.' It applies to every field.

2. Newton' Laws on Motion; "An object in motion remains in motion unless otherwise acted upon by an outside force".

-   As per Chemistry; The Electron remains in motion unless otherwise acted upon by an outside force.

-   As per Biology; The Womb continues to create the baby unless otherwise acted upon by an outside force.

-   As per Time, as all Energy dimensions are represented by Newton' Kinetic Force Laws; Energy remains in motion unless otherwise acted upon by an outside force. The question here is; "What causes the Energy".

When comparing to every other field in regards to the Laws of Thermodynamics, which is a Law of Energy Forces, you see the same trend. The Energy Dimension only exists because Atom/Area exist as a Rule. As per Result? This is (Space/Matter)|(Time). Time is the 3rd Result Option that exists because of a minimum potential fusion between Space/Matter.

Where Einstein made an error was not in recognizing the dual nature of Energy/Time as Chemical and Geometrical. Where he made an error was in his missed assessment that Time itself is then the byproduct Result of the existence of both of the primary fusion options I.E. Space+Matter. Energy/Time exists in Rule/Result because of the original Primary Dichotomy options Area/Atom or Space/Matter. This is expressed via;

Biology with the Womb

Chemistry with the Electron

Geometry pertaining to Breadth

Communication regarding Range

Choice with the maybe option

Etc.

No matter the field, there are 3rd options that only exist due to the primary dichotomy options. This is true in both Rule and Result, even further so to the extent that if you have a Rule and a Result, you will have a Representation between them.

CHAPTER 11: Analyzing Game Theory & ToE

(A Case Study)

Psychology/Philosophy via Game Theory & Equilibriums and the ology of "Rules" compared to "Results".

Gaming: Command & Conquer Red Alert 2; Multiplayer Skirmish

Focus: Interaction between players with a primary comparison on primal nature in an environment with no real world consequences. Secondary Goal: Set parameters between verbal and non-verbal interaction and compare responses.

Summary Data;

      Gaming is a good metric for judging psychology through a philosophical lens as a sort of an objective, albeit abstract case study. Which is not a point against the study. In fact, as all rules are "perfect world", in removing the biases of real life you can actually gauge intent or potential reasoning in the "fair competition" sense, that never truly exists in reality. Abstract reasoning via objectivity in perfect world, fabricated environments is a genuine foundation in Science; To parlay data from one field to another or to compare/contrast data within a field via overlapping and/or transferable data I.E. "The birds and the bees" as opposed to "the humans" (Which, in and of itself is a scientific point to tone down the sexual education from the extremities the educational system has taken them to in my opinion. By my estimation, there is no need for the extremities unless to otherwise propagate that which you're attempting to prevent).

This game particularly is a war game wherein the objective varies based on mission as per single player gameplay. However, in the skirmish environment the basic goals always remain the same, though as part of this claim may be independently contested, it will itself be defended hereafter;

- To build a base.

- To amass an army.

- To defeat your enemies.

The parameters of the game itself change based on preferences to agreed upon rules that are decided in advance including location. As per my data, these game settings are up to the host of the game, with the caveat that if enough people dislike the settings then the host of the game tends to concede to an alteration more often than not, which itself is a piece of psychological data to be considered when weighing the entirety of the study. Also notable is how rare it was for someone to refuse to play by the host' rules even after an objection was made.

My conclusion slowly became that if they cared enough to complain then this was the game that they wanted to play or in the very least, was good enough for them to play it even if not by their own rules, as each person could start their own game or join a game that had already been started by someone else and still chose not to. Those that were opposed to the host' rules due to location, no allowance for "superweapons" or any main setting for the game itself would tend to just exit the lobby and pick a different game without complaint.

With the simple act of staying in the game lobby and complaining, they had already begun to resign themselves to the notion that this would be the game that they chose to play. When accepting that they were attempting to gain leverage at that point, there was a high probability that they would operate within the confines of the rules. Whereas if a single concession was made to anyone? This immediately started a pre-game conflict that would have implications within the game itself, merely via trying to agree upon parameters.

Occasionally, using a 4 player metric, 1-2 of the players would leave wherein a concession was made. The concession itself, or the amount given thereafter, was seemingly irrelevant. The objector and the conceeding host were all that remained. So long as no concessions were made there was a higher probability of being able to start the game without any pre-game conflict. Also important is the factor of TIME. The variables tend to extrapolate.

Excluded were the "Extreme" variables which are essentially anomalies in the form of being angry about a previous game when joining another game started by the same host and even to a higher dimension, that of carrying over a grudge from one day of playing the game to the next day of playing the game. As all of these anomalies were "Random" in the sense of having no direct attachment to the existent game itself, they're excluded as an infinite inconsistency or "anomaly", thereby something that can only be accounted for as something that can not be accounted for. It is a variable that can be transferred from one system to another without intent.

So then, if any concessions were given, it would take a longer amount of time for someone to be able to START a game under any settings. Thereafter another variable was developed compared to that of "Ugly Duckling Syndrome", in the sense that the longer that a hosted game remained in the lobby, the more likely it was that the players would see the game as undesirable. This adding another complication to the desired result to START the game. This creates a bias before the game itself has even been presented and is illogically applied to the host of the game without ever having seen the game itself. The basic settings are as follows;

- Do you start with troops?

- Do you start with money?

- Do you allow super weapons to exist?

*****There is a decision made on quantity for each individual field. When comparing my equilibrium work to that of Jonathon Nash', I try to stress the importance of acknowledging both "Utilitarianism and Existentialism" or both "Individual and Group Identities". As applied in this environment it is the acknowledgement of both the "Perfect World Rules" as per ToE and also the variance in MAXIMUM/MINIMUM POTENTIAL or the "Spectrum of the Imperfect Results" that is Nash's Equilibrium work. Whereas he establishes the perfect world as all rules do, his focus was primarily on how to maximize potential to get the Results closer to the Rules. In my estimation, some of that study includes an unfortunate bit of bias here and there but on the whole is absolutely brilliant*****

As per my extensive data (having far too much free time on my hands at present, hence this effort to study more on game theory and apply more of my work to the already existent rules in mathematics as applied to the real world), these are some of my conclusions;

1. Diversity is inescapable. The amount of intellectual diversity is never more so apparent than when studying psychology via game theory in

perfect world environments, with fair competition established and the greatest desired result achieved.

This via the strength of the individual first and the group second along with my work acknowledging not only that the groupings are all singular byproducts of the individual components (which was not the focus of Nash' work) but also, that that singular grouping is an individual component itself in a higher dimensional grouping.

You are able to discern genuine, raw, true, psychological data that can be applied outside of the "Game" itself. This is the inherent, extrapolative nature of logic. Diversity is itself inescapable due to "If it exists, its equal opposite exists". It is the by proxy end result of existence itself; Extrapolation.

2. Wherein diversity ceases to exist; Life decays. To remove an individual evolutionary leg or option/choice/variance, is to limit the maximum potential of the entire species. Whenever an aspect of the game is removed? An overall maximum group potential is inherently lowered as a result. As per the game itself?

This leads to a more fair environment but is less appealing to some whom play the game. Wherein using Nash' Equilibrium to establish potential, the game becomes less varied as options are removed thereby automatically limiting "Kinetic Reactions" as per Newtonian Physics. In the atomic world this means that you get a smaller explosion. In a world designed for war? This means that you get a less explosive war.

59

3. The "Rules" stay the same even when accounting for dimensional growth. As evolutionary options are added (game settings) you must account for the dimensional growth to a higher state of existence be it war or peace but your basic rules do not change. They apply the same as before in the higher set or grouping. I.E. If the Rule were (If+Then) + (Then/If) = (Thus) you'd be left with the eventual (Thus) = Descriptor.

Whatever this grouping is, it will inevitably end up as 1 (If) in a higher dimensional grouping leading to the exponential growth of "If; (If): Then; (If+Then) ... Etc. The parameter limit is reached at the description of the "Result" and reset to the primary function. I.E. When removing an option or adding an option you would then be stepping into a lower or higher dimensional grouping.

The physics example would be the "Internal/External Quantum" outside the parameters of any "Collective/Infinity". To reference white light, to remove any of the light options is to limit the potential of the light itself, I.E. The collective white light "Result". To alter one of the basic "Rules" of light alters the group collective result of the light itself. The light then changes shade AND shrinks the "Maximum Potential Lumens" causing dimmer light. This is considered "Amperage and Frequency" as Wavelength breakdown on the Radio spectrum which is, in and of itself, a form of communication.

The idea of extrapolation of the logic in dimensional growth is that this is one light beam that is now part of a greater collection of light beams and the logic then repeats.

CHAPTER 12: The Roman Dodecahedron

(Spoiler Alert: They're Jewish)

This is a centuries old, unsolved, mystery. The more that I've studied, the more data that I've found to prove my claims. The Pentagon/Pentagram symbology was confusing but the Jews did use a Pentagon and/or Pentagram throughout history. It wasn't considered a Wiccan symbol until much later. Constantine the Roman Emperor, whom converted to Christianity, actually chose to use the pentagram on his seal and amulet.

This was originally used in the Jewish Faith to represent the "5 Hebrew Books of the Pentateuch". Up until medieval times, even to Christians in Rome, the five points of the pentagram would represent the five wounds of Christ on the Cross instead of the 5 Hebrew books.

Nobody knows why the 117 Roman Dodecahedron exist and they can't make sense of them. The first one was found in the 1700's. They have spent centuries trying to ascertain their roots. My research has led me to the conclusion that it is Jewish.

Let's start with what we can rule out;

1. From an engineering perspective, could this have something to do with a functional system of sorts? Considering all of the holes are a different diameter? No. Not at all. From an engineering perspective, a 12 sided pentagon/pentagram shape is fine as per equilibrium... Until you make all of the holes a different size. Then it's basically useless as per any real world applications. It loses too much strucural integrity to be a balanced, practical system required for engineering. There is also variance between the dodecahedron themselves. They're small ranging from 1.6 - 4.3 inches.

2. We can rule out calligraphy entirely. We already know the tools they had available for the era, this would make no sense, it was not mass produced enough as they are extremely rare and they did not have more of them that suddenly wore down over time, as they're all made of COPPER. If there were more then they would've been melted down and destroyed for other purposes.

Also true, is that there is no consistency as per usage in drawing circles because when you flip the side that you're on to any other side you have to draw the circle using the opposite hand motion because the size of the circles has swapped and the smaller or bigger hole causes you to have to use a different motion to draw a circle. The lack of consistency & inherent hypocrisy in such a tool proves lack of function.

The holes are also inconsistent between the independent dodecahedron themselves. This means that many of them would not be usable for such a purpose and is therefore it is ruled out as a possibility.

3. It is useless as per any astrological application known to the entirety of mankind and we have theories for everything. This includes navigation as there is no conceivable format wherein they would use or need to use 12 and/or only 12 of these circles for any nautical purposes, land travel, etc. as there is no known mapping of any kind done across any of the fields addressing it, that uses anything even slightly resembling this to travel with.

It simply does not function as such. I would apply the same logic to any thought on binoculars or anything telescopic-like as per ocular assistance. The best the Romans had developed were glass spheres filled with water and no public person has made any claims to contest that these were anything to do with anything similar, nor can I find any data to suggest such a thing.

4. There are no historical references to them in any currently known texts across numerous empires that are pertaining to the Romans. So we have no evidence suggesting that these are Roman at all really, other than for them being found in Rome.

Now let's switch our focus to symbolism;

1. The Temple of Solomon was made of copper in the city of copper mines.

- All of the Dodecahedron are made of copper.

2. Solomon' copper Temple took 20 years to build. The scriptures mention this repeatedly.

- There are 20 round nodules stemming from the Dodecahedron.

3. They used to use a Pentagram/Pentagon to represent the 5 books of the Pentateuch.

- The Dodecahedron is composed entirely of Pentagrams/Pentagons.

4. Elijah used 12 different sized stones to represent the 12 TRIBES OF ISRAEL in his sacrifice while challenging followers of Baal.

- The Dodecahedron has 12 different sized circles with 1 in each of the 12 pentagons.

5. The followers of Elijah had dwindled to less than 200 and closer to 100 by the time he returned from his exile.

- There are only 117 discovered Dodecahedrons.

A copper kingdom with a copper Jewish temple that that took 20 years to make, with 12 tribes represented by Elijah' stones, along with 5 Jewish holy books represented during the era of Elijah and his followers by pentagrams/pentagons?

They are not the Roman Dodecahedron.

They are the Jewish Dodecahedron.

I state that with confidence after a substantial amount of time researching to try to find any evidence based claims. I can't stress enough how little has been verifiable previously, to such an extent that the only answer available has been; "They were found in Rome, therefore they are Romans".

Given the scientific nature of this book I wanted to add something a bit more personal to it from my studies. After originally seeing the random article, it finally crossed my path in a youtube video. I took the opportunity to try to solve a "random" problem. With so much of my work being rooted in foundational logic, I wanted to focus for a moment on something less objective and more succeptible to interpretive flaws.

With all that said, I stand by all of my claims included in this book in their etirety, with this chapter on Dodecahedrons being no exception as I come to my conclusions using logic and evidence.

CHAPTER 13: From Here To There. But Why?

(A Brief Evaluation Of The Past, Present and Future)

How did we get here and where are we going? These are the questions that make the world go round. It's useless to try to give credit to everyone involved in this type of work through the years. The same is true of shouting condemnation at those whom have spoken against such theories. As per this book, I intent to let it speak for itself. Loudly.

Many a brush has been put to the canvas that is "The God Equation" as per the inumerable Theories of Everything. Most of which I find to be rooted in such abstract reasoning that there is barely anything tangible to dissect. Of those providing any real content the work is so riddled with flaws that even the brilliant Dr. Kaku was left to resign himself to the explanation that; "It makes sense, we just aren't smart enough to understand it yet". This was pertaining to his own work on Hyperspace. From a scientific perspective it's inexcusable.

Yet, I'm left to be grateful to every one of them. Those in favor and those opposed. Those whom understood the intricacies and those that did not. This equation is a format for literally all things. Without the seeming chaos and diversity of the world around me it would have been nearly impossible to do the work that I've done.

Freedom of information is paramount in Science. Without truth we have nothing. Without access to data we have NOTHING.

There is no denying that in spite of the censorship and seeming disregard for the very things that help us to evolve as a species, it would have been nearly impossible for me to do this work outside of the United States of America, flawed or not. I never claimed my nation to be perfect, simply the best of what we have available today or at least, it used to be.

The chaos and hysteria of the differing opinions between people so diametrically opposed in many cases, was the very thing that allowed me the opportunity to do the field comparisons and intensive study that needed to be done in order to solve the Theory Of Everything. The greatest chaos brought about the most order as I see it and dichotomy logic was maintained even outside of the work itself. Our greatest strenghts are our greatest weaknesses and vice versa.

I have a high comprehension level. I lived a diverse life. I worked numerous positions across numerous fields. I studied the entire way through. Greek and Roman history. Jews, Christians and Muslims. Chemistry, Geometry and Biology articles. True information. False information. Genuine belief to forced reaffirmation. I studied the psychology of people.

Throughout the passed 5 years I have tested myself as well as others. This is, more than likely, the greatest human achievement I will ever accomplish for the species. I did so by walking through hell and refusing to cater to the status quo. I fought and I fought hard. Jonathon Nash' old professor said to him pertaining to Game Theory; "You know this spits in the face of 50 years of mathematics"? He was correct.

All too often members of our society choose ease, repetition and familiarity over fact, evidence and truth. We have focused all too much on how we feel about things instead of the reality of the world around us. The reality for me is simple; This was always going to be difficult, no matter who solved it. There was no "easy way out".

In the world of science? Familiarity can do some serious damage to progress. Evolution, in a sense, is achieved through conflict. This book will no doubt lead to conflict but it should also lead to evolution. Evolution can be chaotic and worrying and, if you need the example, look no further than virology and how quickly things can change from one issue to the next. That is the nature of evolution.

What I intend to do by releasing this book is to reach those whom can be reached. To offer assistance to those willing to take the help. To educate those whom seek to be educated. I have no desire to try to run the world. I've no desire to tell people what to do with the data. My only intent is to ensure that you have RECIEVED the data and that you are able to make your own decisions accordingly.

From a psychological perspective it is up to YOU the READER to decide how you want the world of science to look moving forward. Do you want evolution? Do you want change? Get used to being wrong about things that you used to be certain about. Only through failure do we learn and grow. This Universe is a living, breathing, beautiful thing and it will not hesitate to turn your entire world upside down in an instant. This is why the ToE Equation was always so important.

Imagine a tool that helps you to format any field. A tool that helps you to use compare and contrast methods on a field level that reveals data that might otherwise be too difficult to ascertain. This is a useful tool for anyone, especially a species that wants to travel throughout the Universe. It is, in a very real sense, a software update for our hardware. It's better logic that all of us can use in our lives.

To travel throughout the Universe requires an acceptance that we are only as strong as our weakest link. This also requires an acceptance that we can not allow ourselves to stifle diversity as an effort to force the species into a global conformity. This doesn't remove the lowest common denominator, it caters to it and leads to regression.

The chaos is a byproduct of diversity but it is in a symbiotic relation with the fundamental order of the rules. There is simply option, choice, variant, etc. That then extrapolates in higher dimensional equations. In order to understand things as they are, you need to be able to break them down into simple terms and dissect the basics of a things existence, whatever the subject matter may be. Thus, the road to evolution and maximum potential remains; objectivity.

What I have done is to provide the world with a better tool to objectively assess the world around them down to a field level. With a universal format for the engineering process that functions in every field of existence? Your average person is able to dissect more data and compare it better with other data. We can then understand things that we were not able to understand yesterday because we have a tool that helps us to dissect "new" data.

In short, the goal moving forward should not be to stiffle the individual. If my work has done nothing else it has described to the Nth that; "The Group Is The Byproduct Of The Individual Components And Is Only Then Itself One Individual Component Of A Higher Dimensional Grouping". This is backed up with exacting science and mathematical calculation. It lines up with Jonathon Nash' work on Game Theory and Equilibriums as well.

Where Nash' errors wherever applied is in the presumptious nature of the work. The perfect world rules have to be applied to the reality around us to then reverse engineer a maximum potential in reality that is as close to the perfect world rules as is possible. Nash work presumes that everyone does what is best for themselves. Wherever a person is not doing what is best for themselves, Nash Rules cease to apply.

My work always applies as it is dealing with perfect world rules whereas, in analyzing Nash' work, it appears an attempt to establish perfect world rules that can not be applied with a tangible accuracy. For every variable that is accounted for it becomes a threat which, in the case of war, can just as easily turn into a self-fulfilling prophecy as it's treated as a guide for the inherent actions and reactions.

For every variable that is added in Game Theory, it is following the logic of doing what is best for yourself. Some of us do not do what is best for ourselves. Some of us do what is best for other people, even when it hurts us. Some of us just want an honest game. I don't even care about winning the game so long as I'm telling the truth.

ABOUT THE AUTHOR

Professor JT Springston

"Multi-Field Specialist "Jack Of All Trades. Master Of None"

Born in Grand Forks, North Dakota USA, on May 15th, 1987, to loving parents Edward William Springston & Connie Louise Springston, before moving back to Louisville, Ky with his family before the age of 2.

Recieved the school Math Award in 5th grade, graduating from Eisenhower Elementary and moving on to a Magnet Academy, Conway Middle School, thereafter winning 2 talent shows for singing.

Published at a young age and recipient of 2 Young Authors Awards with a national reading and comprehension level in the 96-97th percentile range.

Politically active, involved in political campaigns at various levels, most notably Ed Springston (Father) and the late, great, Gatewood Galebraith during his last campaign for Governor of Kentucky along with Michael Lewis of Independent, Ky.

A musician at heart with schooling in both Orchestra and Choir. Transitioned from classical style to rock bands and eventually to a career as a Hip Hop & R&B artist. Toured locally as a working artist with local radio and television appearances.

Started working at a young age (14), with a diverse employment history including the fields of construction & demolition, food service, warehousing & assembly, production & management, promotion, computer diagnostics & repair, landscaping, sales, writing, etc.

Without any formal education beyond high school and thus, without a single college degree, this work thereby being independently produced, has remained unrefuted by any national or global institution, for over 4 years, with preliminary and "post-theory" submissions and inquiries ranging from Ivy League Universities to varied, renowned scientific publications throughout the nation.

Had numerous discussions with US Army Ft. Knox either in person, online and over the phone with one of their astrophysicists, Dr. Michael Carini pertaining to patent concepts. Along with numerous other discussions with the patent office, Congressman Massie, professors and even the FBI as more directly related to the Equation. To this day, it remains entirely unrefuted by all investigative parties.

Currently designing new patents, some of which involving fusion concepts, writing books, annoying other smart people & building the foundations of a new school of thought based on the Theory of Everything; "JT's School for the Evolution of Intelligent Design".

Jewish, American, Libertarian-Conservative, Humanist